The Institute of Biology's
Studies in Biology no. 54

The Biology of
Plant Phenolics

J. R. L. Walker

B.Sc., Ph.D., F.N.Z.I.C.
Senior Lecturer in Plant Biochemistry,
Botany Dept., University of Canterbury, Christchurch, New Zealand

Edward Arnold

First published 1975
by Edward Arnold (Publishers) Limited,
25 Hill Street, London W1X 8LL

Boards edition ISBN: 0 7131 2479 2
Paper edition   ISBN: 0 7131 2480 6

40207

First printed in Great Britain by
William Clowes & Sons Limited,
London, Colchester and Beccles

General Preface to the Series

It is no longer possible for one textbook to cover the whole field of Biology and to remain sufficiently up-to-date. At the same time teachers and students at school, college or university need to keep abreast of recent trends and know where significant developments are taking place.

To meet the need for this progressive approach the Institute of Biology has for some years sponsored this series of booklets dealing with subjects specially selected by a panel of editors. The enthusiastic acceptance of the series by teachers and students at school, college and university shows the usefulness of the books in providing a clear and up-to-date coverage of topics, particularly in areas of research and changing views.

Among features of the series are the attention given to methods, the inclusion of a selected list of books for further reading and, wherever possible, suggestions for practical work.

Readers' comments will be welcomed by the authors or the Education Officer of the Institute.

1975

The Institute of Biology,
41 Queens Gate,
London SW7 5HU

Preface

Higher plants secrete many complex organic compounds whose role in the life of the plant may not always be very obvious. Among these secondary metabolites those derived from phenolic precursors have provoked much interest among chemists, biochemists, botanists and food technologists. For example lignin, a phenolic polymer, is a major waste product of the pulp and paper industry, yet is vital in the life of the plant. By contrast the anthocyanin pigments of flowers and fruits are a source of much aesthetic satisfaction and have also provided a fascinating insight into the genetic control of biosynthetic pathways whilst tannins affect the palatability of our food and drink.

The aim of this book is to provide an introduction to this area of plant biochemistry together with a guide to some simple experimental investigations which may be carried out with a minimum of elaborate apparatus.

Christchurch, 1975
J.R.L.W.

Contents

1 Structure and Classification of Plant Phenolics

The term 'phenolic compound' embraces an embarrassingly large array of chemical compounds possessing an aromatic ring bearing one or more hydroxyl groups together with a number of other substituents and even within the confines of the plant kingdom Harborne and Simmonds (1964) have classified plant phenolics into fifteen major groupings. However, this small book will only attempt to deal with the major or common phenolic constituents of plants which, for convenience, will be divided into two main groups:

Phenolic acids and coumarins; (C_6-C_1 and C_6-C_3 structures)
Flavonoid compounds, including anthocyanidins; ($C_6-C_3-C_6$ structures)

The following brief review of the properties of some of the more common of these groups of compounds is intended to serve only as an introduction to the subject; one of the more authoritative reference texts (Geissman, 1962; Harborne, 1964) should be consulted for more detailed information.

1.1 Phenolic acids and coumarins

Two families of phenolic acids are commonly found in plants, a range of substituted benzoic (C_6-C_1) acid derivatives and those derived from cinnamic (C_6-C_3) acid which are very widespread. Both types of phenolic acid usually occur in conjugated or esterified form.

1.1.1 Benzoic acids

R = R′ = H   p-Hydroxybenzoic
R = OH, R′ = H   Protocatechuic
R = OMe, R′ = H   Vanillic
R = R′ = OMe   Syringic

R = H   Salicylic
R = OH Gentisic

The structures of the various benzoic acids that occur in plants are summarized above and in Table 1 whilst their distribution has been investigated by several workers. Three of them, p-hydroxybenzoic, vanillic and syringic acids, occur in vast amounts as combined forms in lignin from

2

Table 1-1 Structural relationships of phenolic compounds. Note the similar

Benzoic acids	*p*-Hydroxybenzoic acid	Protocatechuic acid	Vanillic acid
Cinnamic acids	*p*-Coumaric acid	Caffeic acid	Ferulic acid
Coumarins	Umbelliferone	Aesculetin	Scopoletin
Anthocyanidins	Pelargonidin	Cyanidin	Peonidin
Flavonols	Kaempferol	Quercetin	Isorhamnetin
Flavones	Apigenin	Luteolin	

substitution patterns of the B-rings derived from cinnamic acid

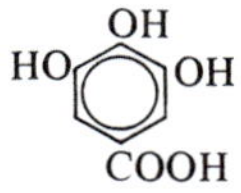

Gallic acid

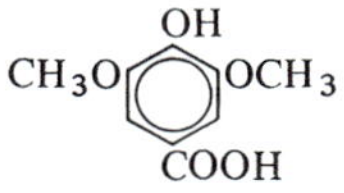

Syringic acid

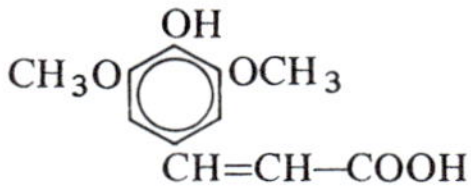

Sinapic acid

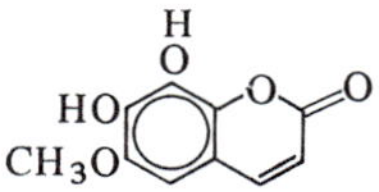

Fraxetin

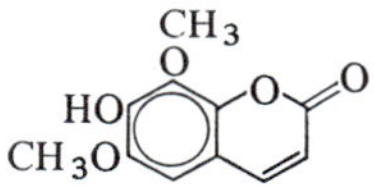

Isofraxetin

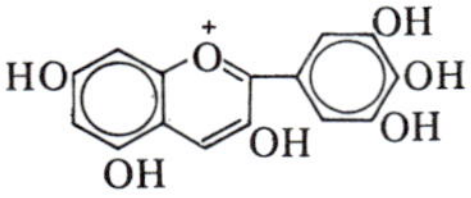

Delphinidin

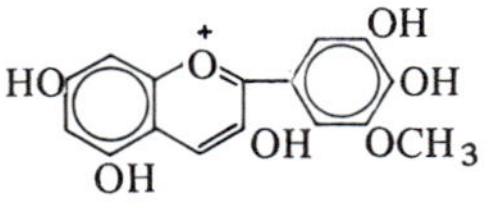

Petunidin

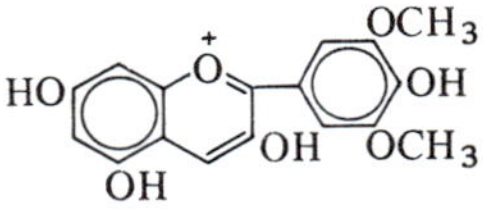

Malvidin

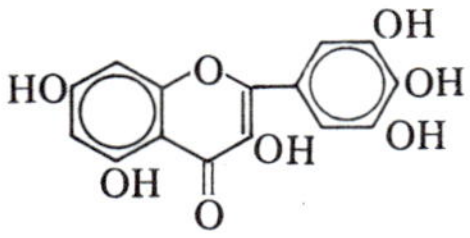

Myricetin

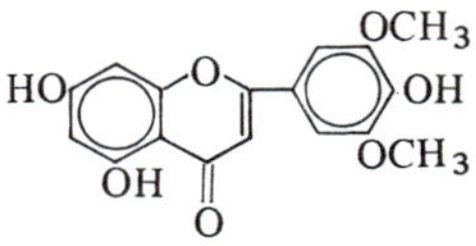

Tricin

which they may be liberated by alkaline hydrolysis. Note that in most plants naturally occurring phenolic compounds have their extra hydroxyl groups adjacent or *ortho* to the C-4 phenolic OH group so that gentisic acid is unusual in this respect yet this acid has been found to occur in 97% of plants examined (Tomaszewski 1960). Protocatechuic acid is similarly widespread but the corresponding tri-hydroxy acid, gallic acid, is only found as a constituent of tannins or as its dimer ellagic acid (see Section 4.2).

The distribution patterns of the various phenolic acids is of considerable taxonomic interest.

1.1.2   Cinnamic acids

R = R′ = H	*p*-Coumaric
R = OH, R′ = H	Caffeic
R = OMe, R′ = H	Ferulic
R = R′ = OMe	Sinapic

Only four cinnamic acids are well known in plants and of these *p*-coumaric and caffeic acids are by far the commonest. However, these acids do not normally occur free but as depsides or esters of quinic or shikimic acid or as sugar esters. One of these depsides, chlorogenic acid (3-caffeoyl-quinic acid) is of almost universal occurrence in plants and is a major substrate for the enzyme diphenol oxidase (ref. Chapters 2, 3, and 6).

The double bond in the side-chain of the cinnamic acids and their derivatives causes them to exist as *cis* and *trans* isomers, the latter being the more stable. These isomers show different biological properties and may be separated by chromatography in aqueous solvent systems such as 5% acetic acid.

1.1.3   Coumarins

R = H	Umbelliferone
R = OH	Aesculetin
R = OMe	Scopoletin

o-Hydroxycinnamic (*o*-Coumaric) acid is rarely found in plants since it readily cyclizes to form coumarin, the aromatic principle of new mown hay and over fifty different hydroxylated coumarins are known. These exist as a variety of sugar esters or glycosides and their distribution is restricted to a few plant families.

From Table 1.1 it will be seen how the substitution patterns of umbelliferone, aesculetin and scopoletin correspond to those of *p*-coumaric, caffeic and ferulic acids. This substitution pattern will be seen again and again throughout the range of plant phenolics.

1.1.4   Phenolic amino acids

The aromatic amino acids phenylalanine and tyrosine occur in plants and animals where they are involved in the biosynthesis of a number of other phenolic compounds (ref. Chapter 2).

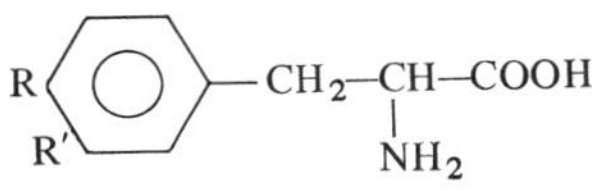

$$R\langle\bigcirc\rangle - CH_2 - \underset{\underset{NH_2}{|}}{CH} - COOH$$

R = R' = H	Phenylalanine
R = OH, R' = H	Tyrosine
R = R' = OH	Dihydroxyphenylalanine (DOPA)

Glycosides of DOPA occur in bean plants and DOPA is also the precursor of the brown melanin pigments of animals.

1.2   Flavonoid compounds

This group of compounds share a basic $C_6 - C_3 - C_6$ structure and includes by far the largest and most diverse range of plant phenolics. In general the flavonoids are soluble in water and show close structural relationships as exemplified in Table 1.1. Flavonoids include the red and blue anthocyanin pigments of flowers, the yellow flavones, and, less common, the aurones, chalcones and isoflavones. These compounds have special interest since they have been used since ancient times to dye cloth and more recently (1915-1940) provided one of the earliest introductions to biochemical genetics and chemical plant taxonomy.

Most flavonoids occur as glycosides in which the $C_6 - C_3 - C_6$ aglycone part of the molecule is esterified with a number of different sugars; however, the present discussion will be chiefly concerned with the various flavonoid aglycones.

1.2.1   Anthocyanidins

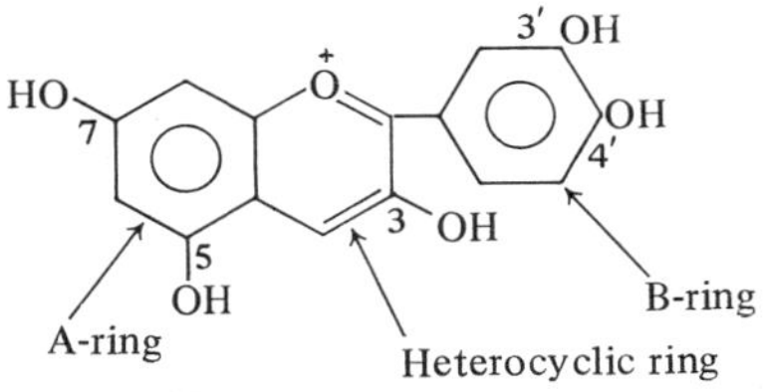

Cyanidin

Anthocyanidins all possess the basic flavylium structure shown above and they normally occur as glycosides in which case they are known as anthocyanins. Six different substitution patterns occur on the B-ring and the effect of this is to modify their colour as shown in Fig. 1-1.

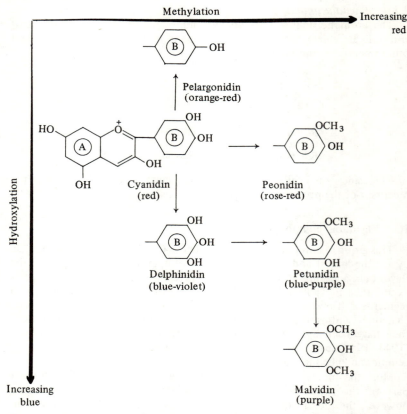

Fig. 1-1   Relationship of anthocyanidin colour to the substitution pattern of the B-ring

Anthocyanidins are widely distributed in the leaves, flowers and fruits of higher plants with cyanidin being the most common. However, the actual colour of a flower is dependent upon a number of factors such as the pH of the cell sap, the presence of certain metal ions and copigmentation with other flavonoids. Anthocyanidins are red in an acid medium becoming blue or purple at alkaline pH; however, they are unstable in alkali due to opening of the heterocyclic ring. For this reason all operations with anthocyanidins are best carried out in dilute acid solution.

If an anthocyanidin possesses *ortho*-dihydroxy groups on its B-ring it may form complexes with iron, aluminium, molybdenum or borate ions and this property is often of great value for identification since the λ_{max} may be significantly altered in the presence of these ions. It is thought that some type of metal chelation is involved in the blue coloration of many flowers but additional factors such as copigmentation with other flavonoids and the formation of anthocyanin-protein complexes may also be involved. For example, both the red and blue flowers of the common *Hydrangea* possess the same anthocyanin pigments but, as every keen gardener knows, their flower colours may be altered by applying traces of iron or aluminium salts to the soil.

At this point is is convenient to mention here the specialized group of red pigments formerly known as the 'nitrogenous anthocyanins' but now more correctly called the betacyanins and typified by the red pigment of the beetroot. Betacyanins are uniquely restricted in distribution being confined to members of the Centrospermae, a group of plants whose classification is still open to controversy so that additional chemical evidence is of great taxonomic importance.

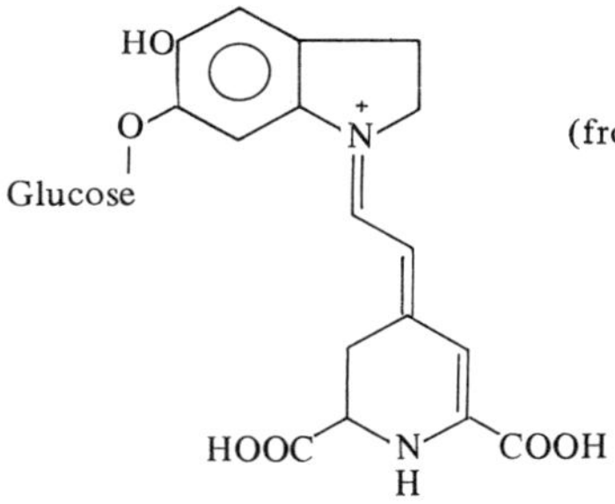

Betanin
(from beetroot)

Betacyanins may be differentiated from anthocyanins by electrophoresis on paper or cellulose thin-layers at pH 4.5 since only the betacyanins are mobile under these conditions.

1.2.2   Flavonols

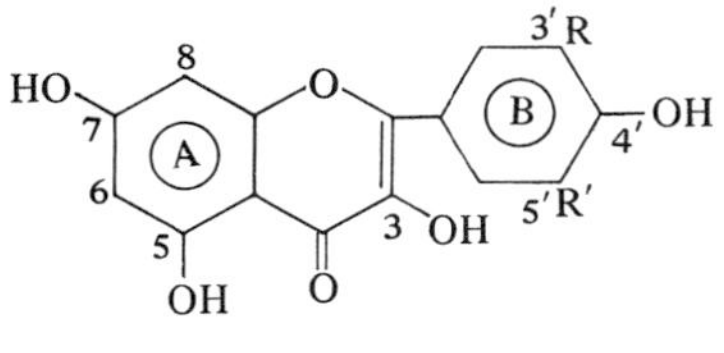

R = R' = H	Kaempferol
R = OH, R' = H	Quercetin
R = R' = OMe	Myricetin

The anthocyanidin pigments of flowers and fruits are almost invariably accompanied by the closely related pale yellow coloured flavonols which also occur frequently in leaves as the following table shows:

Frequency of occurrence of flavonols

Leaves (Swain and Bate-Smith 1962)		*Flowers* (Reznik 1956)
Kaempferol	48%	42%
Quercetin	56%	68%
Myricetin	10%	(not determined)

It should be noted that the basic hydroxylation and glycosylation patterns of the flavonols parallels that of the anthocyanidins. However, minor structural modifications of the A and B-rings frequently occur and these are often of taxonomic importance. For example: iso-rhamnetin (quercetin-3-methyl ether) is common in the pollen of many Graminaceous plants; hydroxylation of the A-ring in the 6 and 8 (less common) positions occurs in quercetagetin and gossypetin which are frequently present in members of the *Compositae* and *Leguminosae*. The latter genus also contains the rarer flavonols fisetin and robinetin which lack the 5-hydroxy group.

1.2.3 *Flavones*

$R = R' = H$ Apigenin
$R = OH, R' = H$ Luteolin
$R = R' = OMe$ Tricin

Flavones differ from flavonols in that they lack the 3-hydroxyl group on the heterocyclic ring but again they show a similar hydroxylation pattern on the B-ring. Only apigenin and luteolin are common in the angiosperms whilst tricin is rare except in grasses (Bate-Smith 1956). This observation is unexpected since the tri-hydroxylated B-ring is otherwise of widespread occurrence for example as sinapic acid in lignin and malvin in flowers.

A number of other flavonoids including flavonones, iso-flavones, catechins and leuco-anthocyanidins occur in plants and their limited distribution makes them of special interest to the chemical plant taxonomist. The formulae and structural inter-relationships of the various plant phenolics are summarized in Table 1-1.

1.3   Phenolic glycosides

Most flavonoid aglycones and many other phenolic acids exist within the plant cell linked to sugars as glycosides. This linkage is usually between a phenolic hydroxyl group and any one of a large array (over 50) of sugars. Glycosidation renders the aglycone more soluble in the cell sap and may also confer stability as is the case with the anthocyanins.

It will be obvious that as a consequence of the multiple hydroxylation patterns of most flavonoids together with the range of sugars found in plants, there may exist a bewildering multiplicity of phenolic glycosides in any one plant; this may be exemplified by apple peel which has been found to contain the following quercetin glycosides; 3-arabinoside,

3-glucoside (isoquercitrin), 3-galactoside (hyperin), 3-xyloside and 3-rhamnoside (quercitrin), by contrast cyanidin-3-galactoside (idaein) is the only anthocyanin present. However, a few general points may be noted:

(a) The sugar/aglycone linkage is usually β except in rhamnosides and arabinosides. Many of these sugars are found only in association with flavonoids.
(b) To add to the complexity the sugar may be acylated (e.g. Tiliroside = quercetin-3-*p*-coumarylglycoside).
(c) The point of attachment of the sugar varies for different types of flavonoid: for example in anthocyanins it is always at the 3-hydroxy position, but 3:5-diglycosides also occur whereas flavonols are usually glycosylated at the 3- or 7-position but rarely at the 5-position. However, B-ring glycosides are also found in some plants.

1.4   Identification of plant phenolics

Very often in biochemistry a new technique has provided the key to open the floodgate of new knowledge and in the case of plant phenolics Bate-Smith pioneered the application of paper chromatography to the study of plant phenolics and chemical plant taxonomy.

Obviously a small book like this cannot hope to deal fully with the numerous and special problems met with during the investigation of phenolic compounds in plants; for this the reader is referred to the comprehensive texts by Harborne (1964, 1973) or Ribereau-Gayon (1972). However, a few general points are of interest since they illustrate some of the special properties of this class of compound.

The first problem in a study of plant phenolics is their extraction from the plant tissue with the minimum of chemical damage such as oxidation. This is frequently made more difficult because many plant tissues contain powerful diphenol oxidases. It is usual therefore to attempt to extract phenolics with water/alcohol or water/acetone mixtures and often a reducing agent, such as cysteine or ascorbic acid, may be added to minimize oxidation. The use of previously dried plant material may help avoid some of the problems. The reader is referred to Section 6.4 for practical details.

The next stage in an investigation of plant phenolics is usually the release of the aglycones by hydrolysis of the glycosidic bonds with dilute HCl or, in some cases, the use of hydrolytic enzymes. Acid hydrolysis may also bring about the condensation of leucoanthocyanidins to yield red 'phlobaphenes' (ref. Section 4.2).

1.4.1   Chromatography of phenolics

After hydrolysis the phenolic aglycones may be extracted into an organic solvent (usually ethyl acetate) and separated by chromatography on paper or cellulose thin-layers. There is an embarrassingly large selection of solvent systems available but the author has found those listed in Section 6.4 adequate for most purposes. Meanwhile a few points of interest with regard to these solvents may be mentioned here:

(i) Benzene, acetic acid, water (BzAc): this is a most valuable solvent since R_F values may be correlated and grouped according to the number of free phenolic hydroxyl groups. Furthermore conjugated phenolics do not move off the origin in this solvent.

(ii) 5% Acetic acid (HAc): aqueous solvent systems separate the *cis*- and *trans*-isomers of the cinnamic acids derivatives giving rise to 'double-spots' (Challice and Williams 1966).

(iii) 'Forestal' solvent: this is the solvent of choice for anthocyanidins (see Table 6-2).

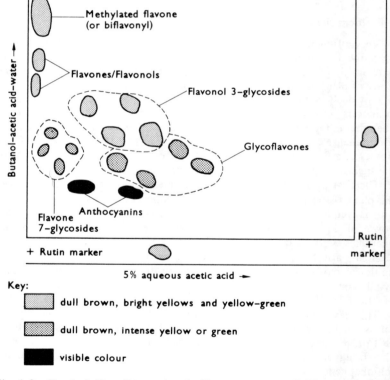

Fig. 1-2 Typical Two-Dimensional Chromatogram of Flavonoids in a Plant Extract. The above chromatogram indicates approximate position common flavonoids take up on a chromatogram. During extraction, some hydrolysis of glycosides may occur, so that small amounts of flavone and flavonol aglycones are usually found. (From Harborne, 1973)

The separated phenolics may be located by viewing the chromatogram under UV light before and after exposure to ammonia fumes which often changes the colour of their fluorescence. Many phenolics give characteristic colours when the chromatograms are sprayed with diazotized *p*-nitraniline or sulphanilic acid. The identity of the sugars may likewise be elucidated by standard chromatographic procedures. Once a suitable chromatographic separation has been found, it may be scaled up to permit isolation of small quantities of the various compounds.

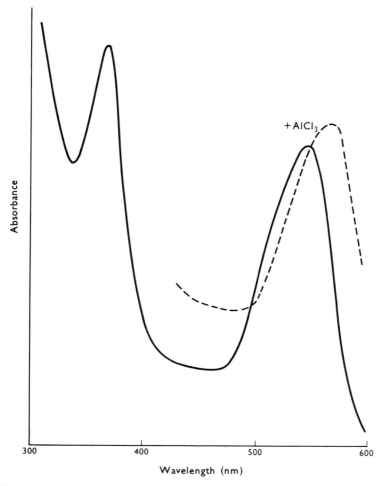

Fig. 1-3 Absorption spectrum of cyanidin in methanolic HCl. The dotted line shows the bathochromic shift due to the addition of AlCl$_3$

1.4.2   Spectroscopy of phenolics

The phenolics and flavonoids having been separated by chromatography, the individual components may be eluted with water or methanol or ethanol and their absorption spectra recorded. The UV and/or visible spectra of phenolics provides a valuable aid to their identification since the wavelengths of the absorption maxima are often characteristic of a particular class of compounds. Moreover the absorption spectra may be changed by the addition of selected reagents. For example, alkali (as sodium ethylate) will ionize the phenolic hydroxyl groups with consequent spectral changes. Similarly the addition of $AlCl_3$ may be used to detect conjugated side-chains (see Fig. 1-3).

Ortho-dihydroxy groups may be identified by treatment with Na acetate/borate which brings about a bathochromic shift of the absorption maximum. The various anthocyanidins all possess characteristic spectra but the wavelength of the maxima are different for aqueous and alcoholic solvents.

Finally it must be emphasized that the safest and surest way to identify a flavonoid is by a comparison of R_F values and spectra to those of an authentic specimen of the compound examined under the same conditions.

2 Biosynthesis and Metabolism of Phenolic Compounds

Neish and other workers have demonstrated that the two aromatic rings of flavonoids arise from two different, metabolic pathways. The C_6-C_3 phenyl-propane moiety is formed via the shikimic acid pathway whilst the A-ring arises from the condensation of three 'acetate' units. These two pathways will be considered separately.

2.1 The shikimic acid pathway

This important metabolic pathway was first elucidated by Davis (1955) who was investigating the biosynthesis of aromatic amino acids in micro-organisms when he isolated biochemical mutants lacking the ability to make aromatic amino acids but whose growth could be restored if shikimic acid was supplied; he therefore concluded that shikimic acid might be a common precursor for the biosynthesis of these compounds. Subsequent studies by many other workers have established that the shikimic acid pathway, shown in Fig. 2-1, is the principal route for the production of such vital compounds as phenylalanine and tyrosine both in plants and animals.

Basically the shikimic acid pathway involves the initial condensation of phospho-enolpyruvic acid from the glycolytic pathway with erythrose-4-phosphate, an intermediate from the pentose-shunt pathway. A series of reactions leads to shikimic acid which is then phosphorylated and combines with a second molecule of phospho-enolpyruvate to yield prephenic acid via the recently discovered chorismic acid. Prephenic acid then loses CO_2 to form phenylpyruvic acid or p-hydroxyphenylpyruvic acid which subsequently undergo transamination to yield phenylalanine or tyrosine respectively.

In animals, but not in plants, the benzene ring of pheylalanine may be hydroxylated to form tyrosine which accounts for the non-essentiality of tyrosine.

It should also be noted that although quinic acid is a very common plant acid, it does not lie directly on the shikimic acid pathway. Protocatechuic and gallic acids can also be formed after the enolization of 5-dehydroshikimic acid.

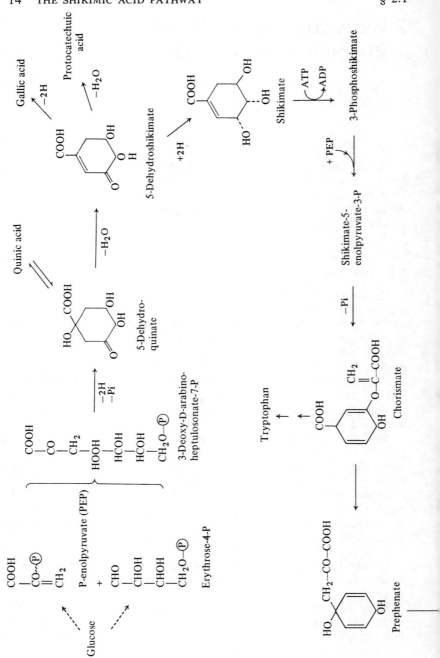

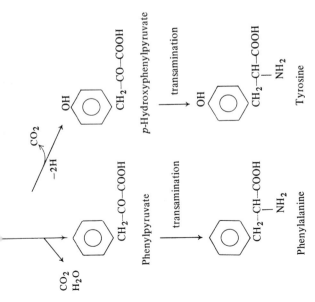

Fig. 2-1 The shikimic acid pathway. ATP is used as an energy source and NADH or NADPH as a source of reducing power

2.2   Deamination of phenylalanine and tyrosine

Many plants have been found to contain a reductive deaminase, phenylalanine ammonia lyase (PAL), which catalyses the irreversible formation of *trans*-cinnamic acid from phenylalanine whilst a similar

enzyme, tyrosine ammonia lyase (TAL), or tyrase, deaminates tyrosine to yield *p*-coumaric acid. Phenylalanine ammonia lyase has been shown to operate reversibly *in vitro* but tracer studies have failed to demonstrate the *in vivo* biosynthesis of phenylalanine from cinnamic acid. Both PAL and TAL are inhibited by the end-products of their reaction.

These enzymes are currently the subject of much detailed investigation because they appear to play a central role in the secondary metabolism of plants; in particular they seem to be involved in the regulation of the biosynthetic pathways leading to flavonoids and lignin.

PAL has been found in all green plants examined so far and also in Basidiomycetes and *Streptomyces* whereas TAL has been found only in grasses and cereal plants. This restricted distribution of TAL agrees with the observation that ^{14}C-labelled tyrosine was a good precursor of lignin when fed to Graminaceous plants whereas ^{14}C-phenylalanine was readily incorporated by all higher plants. Moreover in higher plants high PAL activity is usually found only in those tissues that are actively synthesizing lignin and is almost undetectable in tissues not undergoing lignification.

Much attention has been focused on PAL because this enzyme shows striking changes in activity in different tissues in response to a wide variety of external factors such as light, disease, and growth factors. Changes in external physiological factors often affect lignification and/or flavonoid formation in plants so it seems reasonable to assume that PAL may be involved in controlling these changes. For example, light stimulates the formation of PAL in many tissues and there is some evidence that this may be a phytochrome mediated response in many, but not all, of the systems so far investigated. Likewise wounding and the growth hormone ethylene also stimulate PAL activity in pea seedlings, sweet potato and other plants. The recent review by Camm and Towers (1973) is recommended for a more detailed survey of this important topic.

Once formed cinnamic and *p*-coumaric acids do not accumulate as such but rapidly become conjugated with other compounds such as coenzyme A, quinic or shikimic acids or various sugars before eventually being incorporated into more complex phenolics such as flavonoids or lignin. Presently available evidence suggests that further hydroxylation and methylation of the aromatic ring occurs after the cinnamic or *p*-coumaric acids have been converted to the conjugated form. However, further discussion of this point will be deferred until later sections.

2.3   Formation of C_6–C_3–C_6 flavonoids

Evidence for the biosynthetic routes leading to the flavonoids has come from a number of different experimental approaches including genetic studies, and the comparison of closely related chemical structures, but the major contribution has come from tracer studies.

By feeding plants with ^{14}C-labelled precursors Neish and other workers were able to show that the A-ring of flavonoids was formed by the

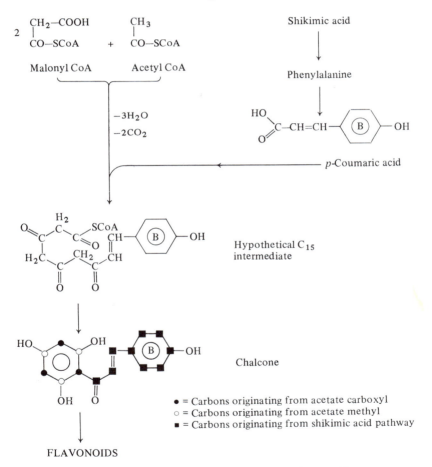

Fig. 2-2   Proposed biosynthetic route to flavonoids

head-to-tail condensation of acetyl-CoA with two molecules of malonyl-CoA to yield a phloroglucinol structure; this is basically similar to the pathway used for the biosynthesis of aromatic compounds in fungi. The remaining C_6-C_3 phenylpropane unit is derived entirely from the shikimic acid pathway and it is thought that cinnamyl-CoA or p-coumaryl-CoA are the immediate C_6-C_3 precursors. These ideas are summarized in Fig. 2-2.

There is still some doubt as to the exact nature of the first formed C_{15}

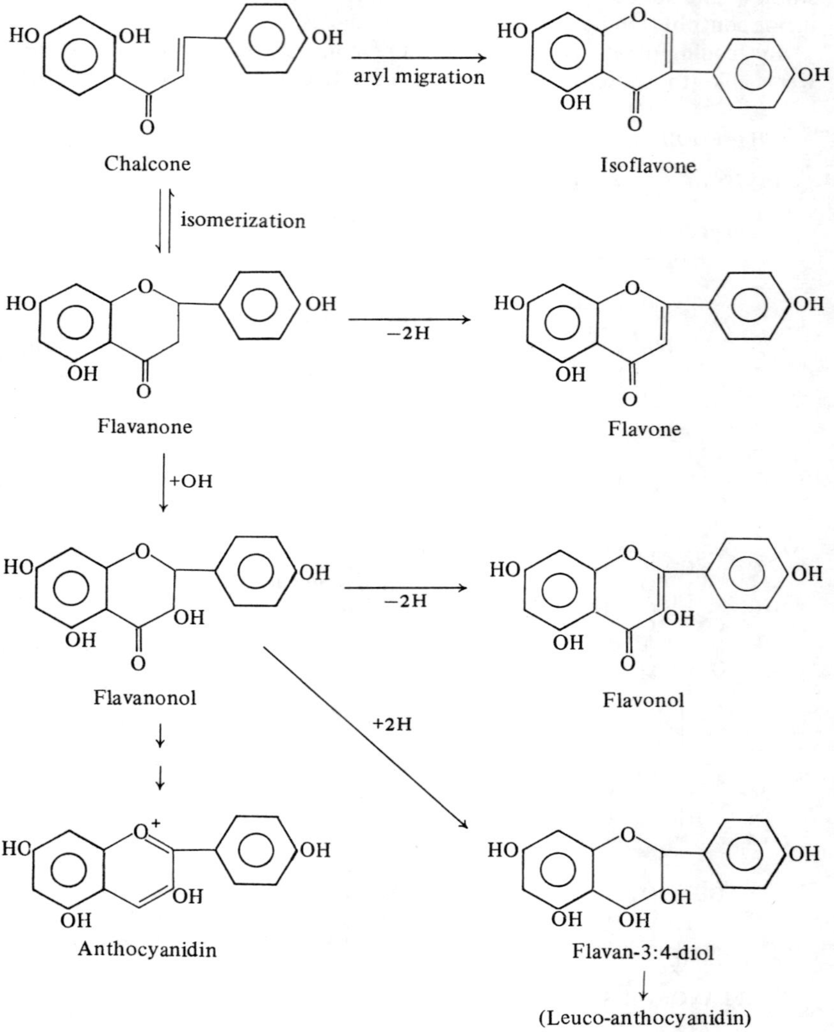

Fig. 2-3   Biosynthesis of the basic flavonoid structures from chalcones

intermediate, but the available evidence supports the view that it is a chalcone since several workers (Grisebach 1967, Wong 1968) have shown that labelled chalcones are readily converted into other flavonoids. Figure 2-3 outlines a scheme, based on chemical and genetic evidence, for the interconversion of the various classes of flavonoids.

2.4   Hydroxylation and methylation of phenolics

Preceding sections of this book have shown that the chief structural differences between members of a particular class of plant phenolics depends upon the degree of hydroxylation and methylation of the phenyl-propane moiety, but in addition different patterns of A-ring hydroxylation also occur. Grisebach (1965) has provided evidence that the hydroxylation pattern of the A-ring (i.e. a basic phloroglucinol or resorcinol structure) was established before the cyclization step and that subsequent modification was virtually impossible.

The exact details of the hydroxylation sequence of the B-ring are still a subject for investigation, but it is generally accepted that the *para* (4)-OH group is introduced at the cinnamic acid (C_6-C_3) precursor stage before the formation of a C_{15} compound. However, any subsequent hydroxylation and methylation reactions involve C_6-C_3-C_6 compounds; this is exemplified by the observations of Underhill *et al.* (1957) that *p*-coumaric acid (4-OH) was a better precursor of quercetin (3':4'-diOH) biosynthesis in buckwheat than was caffeic acid (3:4-diOH).

Hydroxyl groups are introduced into the aromatic nucleus by the phenolase enzyme complex (also called diphenol oxidase, poly-phenol oxidase or tyrosinase) which appears to exhibit both mono-phenol hydroxylase (at the 3-position) and *o*-diphenol oxidase activity. For example Levy and Zucker (1960) investigated the biosynthesis of chloro-genic acid in potatoes and suggested the following scheme:

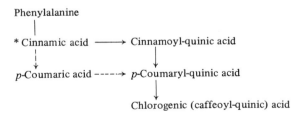

* Probably involved as cinnamoyl CoA.

Other workers (Walker 1964, Sato 1966) have shown that chloroplast preparations containing phenolases could hydroxylate *p*-coumaric acid to caffeic acid.

More recently Kojima and Uritani (1972) studied the biosynthesis of chlorogenic acids in sweet potato and they presented evidence that isotopically-labelled *trans*-cinnamic acid, formed by the action of phenyl-alanine deaminase, became conjugated with a sugar before any subsequent hydroxylation steps to *p*-coumaric and caffeic acid derivatives. Formation of chlorogenic acid and other quinic acid depsides occurred after this stage.

Methylation of the B-ring hydroxyl groups may be effected by *o*-methyl transferases which use S-adenosyl-methionine as CH_3-donor and Finkle and Nelson (1963) have isolated these enzymes from higher plants.

2.5   The oxidation of phenolics: phenolase and laccase

Most higher plants and many Basidiomycete fungi possess enzymes capable of oxidizing dihydroxyphenols. In higher plants these enzymes are usually, but not exclusively, *o*-diphenol oxidases whereas *p*-diphenol oxidases (or 'laccases' so-called because they were first discovered in the sap of the Japanese lac tree) are more common in the higher fungi. Both enzymes require copper as a prothetic group and show many similarities in their properties.

The actual role of phenolase in the plant is not yet clear but available evidence suggests that it is involved in not only the *o*-hydroxylations described earlier but also in enzymic browning and the plant's defence reactions (ref. Section 3.4). As far as plant biochemists are concerned the almost universal occurrence in plant tissues of phenolase together with its substrates forces them to take stringent precautions to avoid enzymic browning and consequent enzyme inactivation when attempting to investigate the biochemical processes of many plant tissues. For example, it is customary to add reducing agents such as ascorbic acid or cysteine or a phenolase inhibitor like polyvinylpyrrolidone (PVP) during isolation procedures for plant enzymes. It is interesting to speculate how the intact plant cell avoids interaction between phenolase and its substrate since they both occur in the same cell!

Laccase or *p*-diphenol oxidase differs from *o*-diphenol oxidase in that it does not exhibit any hydroxylase activity towards monophenols. It is also able to oxidize quinol and *p*-phenylene diamine; unlike *o*-diphenol oxidase it is not inhibited by PVP. There is some evidence that laccase may be involved in lignin biosynthesis (Section 3.5).

2.6   The formation of phenolic glycosides

Most flavonoids occur in living plant tissues as glycosides but little is known as to how and when the sugars are added to the flavonoid aglycones; however, a few general points may be noted. For example, when flavonols and anthocyanins co-exist in the same flower they often exhibit different patterns of glycosylation which suggests that different enzyme systems are involved in their glycosylation and this is supported by genetic evidence. In the case of the anthocyanins glycosylation of the 3-position of the heterocylic ring must occur at an early stage in order to stabilize the otherwise unstable anthocyanidin.

Conn (1964) and other workers have presented evidence that sugar nucleotides, such as UDP-glucose, are involved in glycoside formation and that the sugars are added by stepwise transfer.

2.7   Boron nutrition and plant phenolics

Trace quantities of boron are essential for healthy plant growth but as yet plant physiologists are still unsure of the exact role of this vital nutrient. One manifestation of boron deficiency is an accumulation of phenolic acids which eventually leads to the death of the plant cell. In 1967 Lee and Aronoff presented evidence that the borate ion controlled the partitioning of glucose metabolism between glycolysis and the pentose phosphate shunt pathway. This modulating effect of borate was brought about because the borate ion formed a complex with 6-phosphogluconate and this 6-PG/borate complex inhibited the action of 6-phosphogluconate dehydrogenase, a key enzyme in the pentose phosphate shunt pathway. The level of activity of this pathway affects the supply of erythrose-

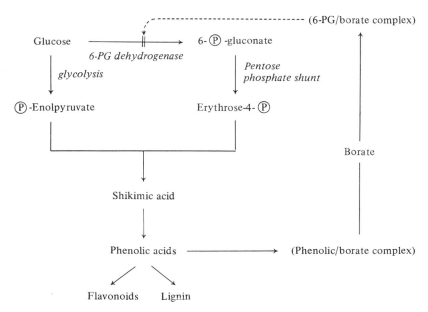

Fig. 2-4   Diagram to show the biochemical effects of boron deficiency (based on data of Lee and Aronoff, 1967)

4-phosphate which, as we have seen already, combines with phospho-enolpyruvate to enter the shikimic acid pathway (ref. Fig. 2-1) to yield phenylalanine and eventually phenolic acids, flavonoids, etc.

In addition certain dihydroxy-phenolic acids also form strong complexes with borate thus causing a further diminution of the amount of borate available to complex with 6-phosphogluconate. This removal of borate from the 6-PG/borate complex results in a release of the inhibition of 6-phosphogluconate dehydrogenase and the subsequent formation of excess phenolic compounds to complex what little borate remains. Thus, in boron-deficient plants, an autocatalytic or positive feedback system is set up which generates an excess of phenolic acids leading to tissue necrosis and the eventual death of the plant.

It is also interesting to note that Rajaratnam et al. (1971) found a complete absence of leuco-anthocyanins in boron-deficient oil palm seedlings; a result in keeping with the above hypothesis.

3 Functional Aspects of Phenolic Compounds in Plants

Phenolic compounds probably constitute one of the most widespread and diverse groups of secondary plant metabolites yet even today many aspects of their role in the life of the plants are still a matter of conjecture as is also their evolutionary significance. This chapter will discuss briefly the structure and roles of certain important groups of phenolic compounds that are found in plants.

3.1 Anthocyanin and flavonoid pigments

The beautiful colours of many plants are not only a source of aesthetic pleasure but also of great scientific interest since studies of the red and blue pigments of flowers have contributed much to our knowledge of biochemical genetics and the control of the pathways of flavonoid biosynthesis. Studies of the distribution patterns of flavonoids in leaves and flowers have opened up a fertile field for the application of chemical analysis to help solve problems in plant taxonomy.

The flavonoid pigments appear to represent a dead-end of phenyl-propanoid metabolism and many possible roles have been suggested for them including those of excretory waste products and defence against invading pathogens, but the only function which seems at all certain is that of attracting pollinating agents. Even this function must have been a secondary development since anthocyanins also occur in leaves, stems and other organs not involved in pollination. Anthocyanins responsible for fruit colour are of considerable economic importance in influencing their consumer appeal and the preservation of fruit colour can create considerable problems for food processors.

Studies of flower colour have provided much of our knowledge of the genetics of flavonoid biosynthesis. In the early 1930s Muriel Onslow and a group of geneticists at the John Innes Horticultural Institute, near London, investigated the pigments of snapdragon (*Antirrhinum majus*) and were able to show that the various simple biochemical events in the biosynthetic pathway were controlled by Mendelian factors. This work also revealed certain basic patterns of inheritance in the flavonoid pigments of many different plant species. However, it was not until Bate-Smith applied the then new technique of paper chromatography to these problems that major advances became possible.

An excellent example of the application of chromatography to studies of the inheritance of flower pigments is given in Table 3-1 which summarizes the distribution of flavonoid pigments in the petals of mutants of *Antirrhinum majus* (snapdragon) flowers.

Table 3-1   Flower pigments of *Antirrhinum majus* mutants*

Genotype	Phenotype	Pigments			
		Cyanidin Quercetin	Pelargonidin Kaempferol	Luteolin	Apigenin
PPMMYY	magenta	+		+	+
PPMMyy	red-orange	+		+	+
PPmmYY	pink		+		+
PPmmyy	yellow-orange		+		+
ppMMYY	ivory			+	+
ppMMyy	yellow			+	+
ppmmYY	ivory				+
ppmmyy	ivory				+

* All strains are genotype NN.

Swain and Bate-Smith (1962) have used these results to suggest the pathway shown in Fig. 3-1 for the biosynthesis of flavonoid pigments. They suggest that four principal genes are concerned:

(i) Gene N which appears to control a very early stage in flavonoid formation since nn homozygotes are albino, do not contain flavo-noids, but do accumulate C_6–C_3 cinnamic acid derivatives.

(ii) Gene Y (yellow) controls the closure of the chroman ring of the chalkone to yield a flavonoid precursor. This results in lower levels of aurones in YY genotypes.

(iii) Gene M (modifier) affects the introduction of the 3′-hydroxyl group into the B-ring of all flavonoids except the aurones. Di-hydroxylation of the B-ring is found in all or none of the flavonoids which implies that gene M must intervene before gene P.

(iv) Gene P (pink) controls hydroxylation at the 3-position of the heterocyclic ring and therefore the formation of anthocyanidins and flavonols in PP and pp genotypes.

In recent years detailed chromatographic analysis of the flavonoid compounds in the leaves and flowers of the different species within a genus has provided a valuable aid to the taxonomist and has often provided useful clues to phylogeny, the origin and evolution of taxa. In the case of the flavonoids there appears to be a general evolutionary trend towards increasingly complex structures. The less highly developed plant orders, such as the gymnosperms, ferns and mosses possess only simple flavonoids, whereas highly evolved families like the Leguminosae and Compositae exhibit a wealth of highly substituted flavonoids.

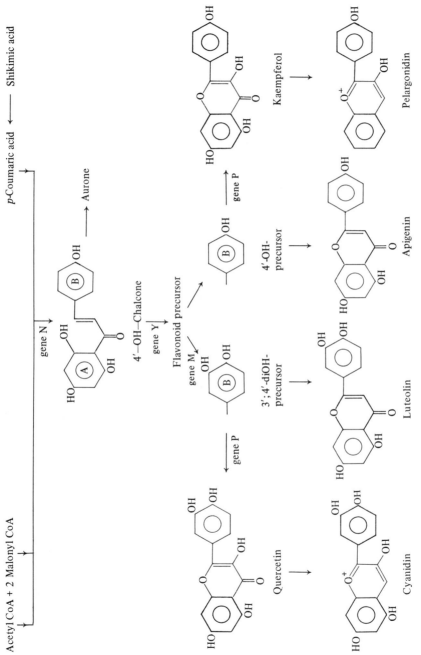

Fig. 3-1 Biosynthetic pathway of flavonoids in *Antirrhinum majus*

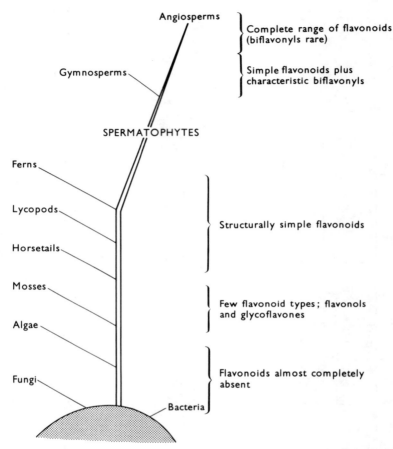

Fig. 3-2   Phylogeny and evolution of flavonoids in plants (after Harborne, 1967. Reproduced by kind permission of Dr J. B. Harborne and Academic Press Inc.)

One other example of chemosystematic studies of anthocyanins will be given. Chromatographic analysis of the flower pigments of various species of *Impatiens* was able to prove conclusively that the orange-flowered *I. aurantica* found in the Celebes Islands was a separate species from the magenta-coloured *I. platypetala* found in nearby Java, the flowers containing aurantidin and malvidin respectively. Many other examples of the use of the chemical approach in helping to resolve taxonomic problems will be found in the references whilst suggestions for simple experimental work are given in Chapter 6.

3.2   Phenolic compounds and our food and drink

The bright colours of many fruits such as plums and strawberries, etc. are due to anthocyanin pigments, but these compounds appear to contribute little to flavour. Nevertheless they are of considerable importance in food technology and wine making where the final colour of the product is important. Many fruits and vegetables also contain considerable quantities of other phenolic compounds such as the chlorogenic acids and tannins which may markedly affect the flavour and astringency of the fruit. A good example of this is given by the differences between cider and dessert apples: the former are rich in phenolics whilst cider made from ordinary apples lacks body and is insipid. Hops similarly contain large amounts of tannins and flavonoid compounds, but their contribution to beer flavour is still a matter of controversy among brewing chemists. However, there seems little doubt that phenolic compounds are involved in the formation of undesirable 'hazes' in beer and wines. The tannins present in tea appear to be related to the tea-tasters' assessment of quality.

Derivatives of cinnamic acid occur frequently in many fruits and vegetables, the commonest being chlorogenic acid. This compound, with catechin, serves as the principal substrate for the phenolase complex which is responsible for the enzymic browning exhibited by many fruits and vegetables such as apples and potatoes. The occurrence of this enzymic browning is disadvantageous in the majority of food processing operations, and steps are taken to either inhibit or destroy the phenolase system of the materials being processed. For example ascorbic acid (vitamin C) is added as an antioxidant during the manufacture of apple juice to prevent unwanted browning.

By contrast in the processing of tea and cocoa or during the manufacturing of cider and perry the enzyme-catalysed oxidation of the original phenolics contributes much to the desirable qualities of the products. Plant phenolics are therefore something of a mixed blessing to the food technologist.

3.3   Pharmacological aspects of plant phenolics

Although not strictly within the definition of the title of this chapter, it is nevertheless of interest to consider briefly the effects of some plant phenolics upon the animal organism and despite the numerous advertising claims for the therapeutic value of 'bioflavonoids' these have not been well-substantiated by clinical trials so will not be considered further. It is fortunate indeed that the vast majority of plant phenolics are non-toxic to animals but certain furano-coumarins (e.g. rotenone) have been used as fish-poisons whilst coumarin is an effective inhibitor of seed germination.

Phloridzin causes glucosuria, a form of diabetes, in animals by interfering with the absorption of glucose in the kidney and small intestine and is much used in physiological studies. Fortunately for man it only occurs in the bark of apple trees and not in the fruit.

Phloridzin
(a dihydrochalcone glucoside)

glucose

Of greater pharmalogical importance are the iso-flavones and coumarins. The former possess weak oestrogenic activity and cause reduced fertility in livestock which have fed on some strains of sub-terranean clover. However, steps are being taken to develop clover strains lacking these iso-flavones—thus the biochemist helps the farmer. Spoiled sweet clover also gives rise to agricultural problems since spoilage causes two molecules of coumarin (a characteristic constituent) to combine to yield a molecule of dicoumarol which is an haemorrhagic factor and is used medicinally to prevent blood clotting.

microbial action

Coumarin   Dicoumarol

Coumarins are also of interest to agricultural scientists because despite their sweet smell they have a bitter taste which sheep find objection-able. Thus the presence of quite small amounts of coumarin in herbage may bring about a marked reduction in the animal's food intake. Luckily for New Zealand's farmers their sheep seem to find the predominant ryegrass-clover pastures most acceptable but this may be the fortunate result of a number of factors affecting herbage palatability.

3.4   Defence mechanisms involving phenolics

The antimicrobial and antifungal activity of phenolic compounds has been known since the days of Pasteur and Lord Lister, so it is not surprising that plant physiologists seeking to find an explanation for the existence of the wide range of plant phenolics suspected that they might be involved in defence against invading phytopathogens but even today this problem is still controversial.

The earliest and best example of disease resistance associated with the phenolic constituents of plant tissues was discovered in 1923 by J. C. Walker and his group, who observed that the outer neck scales of coloured onion bulbs were resistant to infection by *Colletotrichum circinans* whereas unpigmented varieties lacked resistance. This resistance was subsequently correlated with the presence of 3:4-dihydroxybenzoic (proto-catechuic) acid in the scales. Another example is given by the rust resistance of 'Khapli' wheat which is associated with a high level of phenolics. Studies of the phenolic compounds and phenolase enzymes in apple leaves that are resistant to apple scab (*Venturia inaequalis*) have shown that resistance is related to the levels of oxidized phenols. This and other similar observations suggest that it is in fact the quinones, produced by the action of phenolase, that are responsible for disease resistance. These *o*-quinones are highly reactive and are known to combine with amino acids and proteins, in this case those of the pathogen, thus rendering them biologically inactive.

o-Dihydroxyphenol → *o*-Quinone → Brown oxidation products; protein (e.g. extracellular enzymes from pathogen)

Additional support for this hypothesis comes from Japanese workers who found an increased biosynthesis of phenolic compounds in sweet potatoes when they became infected with black root rot.

Perhaps the most exciting discoveries concerned with the biochemistry of plant disease have been those arising from the Phytoalexin Theory of Plant Disease first put forward in 1940 by Müller and Börger. Major experimental evidence in support of this hypothesis came from work in Australia by Cruickshank and Perrin who developed their elegant 'drop-diffusate' technique for the isolation of phytoalexins. These workers inoculated the seed cavities of immature pea or bean pods with drops of a suspension of fungal spores and after incubation for 24–48 hours they were able to isolate various fungitoxic compounds, or phytoalexins, which had diffused from the inoculated host plant's cells. Different plant species produce different inhibitor substances and a list of them is given in Table 3-1.

The results of this work have been formulated into the Phytoalexin Theory of Disease Resistance in Plants. Briefly this theory states that phytoalexins are absent from healthy plant tissues, but they are produced whenever the plant is invaded by a fungus and if produced in high enough concentration they are toxic to the invader. Thus nonpathogens induce a

Table 3-1   Examples of phytoalexins produced by plants in response to fungal infection

Pisatin

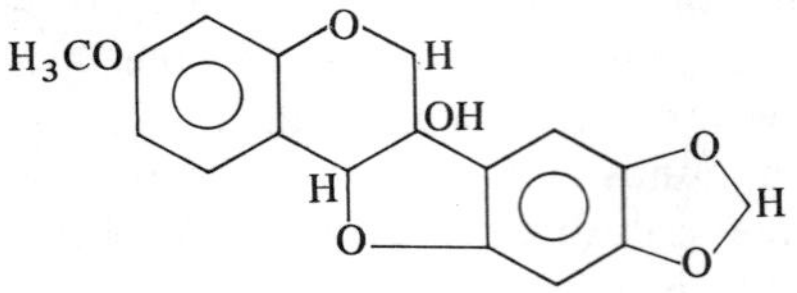

Pea
(*Pisum sativum*)

Phaseollin

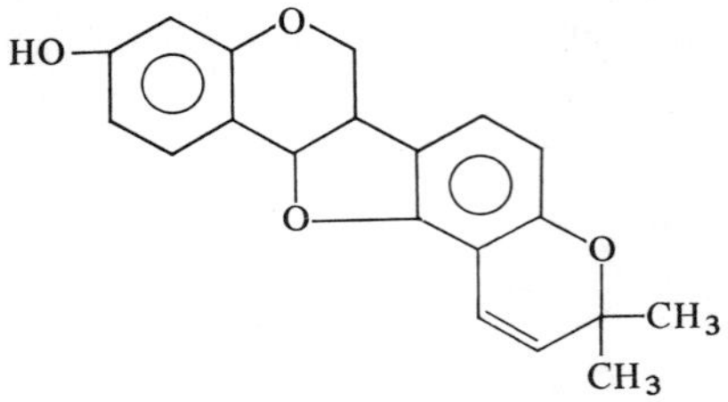

French bean
(*Phaseolus vulgaris*)

6-Methoxy-mellein
(iso-Coumarin)

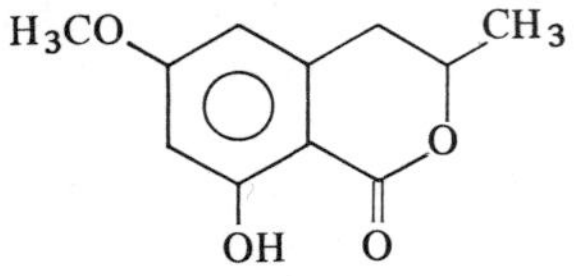

Carrot
(*Daucus carota*)

Orchinol

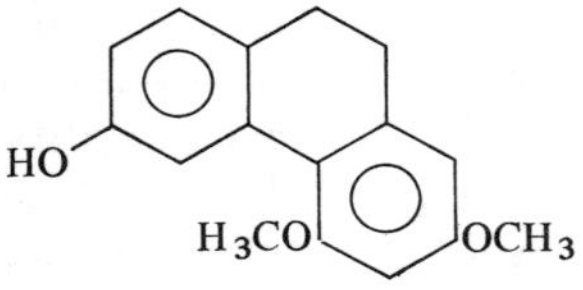

Orchid
(*Orchis militaris*)

Ipomeamarone

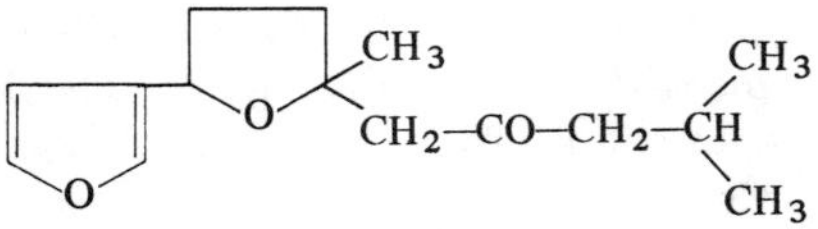

Sweet potato
(*Ipomea batatas*)

lethal level of phytoalexin whilst pathogens are able to grow in the plant because they do not stimulate the biosynthesis of phytoalexin. In other words pathogens are those organisms which do not trigger the defence mechanism. It is important to realize that the resistant state is not inherited but appears as a result of infection. However, the sensitivity and subsequent speed of response of the plant to the invader is genetically controlled. This area of research is a rapidly advancing one and should lead to new methods of biological control of plant disease.

3.5   Phytotoxic phenols: ecological aspects of phenolics

A phytotoxin may be loosely defined as any substance produced by higher plants which is capable of suppressing the growth of other plants. Many such phytotoxic compounds have been discovered and they are frequently phenolic or terpenoid in nature; however, the present discussion must be limited to the former group (a useful reference is Müller and Chou, 1972).

Phytotoxic phenolics have been extracted from seeds, fleshy fruits and from soils containing decomposing plant materials. McCalla and his co-workers have shown that various soils contained phenolic acids including p-hydroxybenzoic, vanillic, p-coumaric and ferulic acids and these compounds can act as germination inhibitors. A similar range of phenolic acids was extracted from cereal residues and this is of considerable significance with respect to the system of stubble-mulch farming practised in parts of the U.S.A.

Müller and his group have made extensive ecological studies of the characteristic 'chaparral' vegetation of Southern California and have suggested that the dominant shrub species secreted phytotoxins which enforced dormancy upon the seeds of other herbage plants. Chromatographic analyses of aqueous extracts of the leaves of the dominant shrub *Adenostoma fasciculatum* revealed the presence of phenolic acids, phloridzin, arbutin and umbelliferone. Rain leaches these phytotoxins from the leaves into the soil thus bringing about a reduction in the rate of germination of seeds in the soil around those plants. A similar pattern of events has been observed with leachates from the litter of *Eucalyptus* species.

Salicylic acid

Juglone

Other trees and shrubs also secrete phenolic phytotoxins; for example salicylic acid released from the leaves of *Quercus falcata* var *pagodaefolia* causes a total suppression of shrub growth beneath the tree. Likewise decomposing leaves from the black walnut (*Juglans nigra*) release juglone which interferes with the growth of surrounding plants.

4 Polymeric Plant Phenolics; Lignin and Tannins

This chapter may be considered as an extension of the previous one since lignin and tannins have functional roles in the plant. However, because these important materials are complex polymers, it is felt that they should be considered separately.

4.1 Lignin

Lignin is a complex three dimensional polymer of phenylpropanoid (C_6-C_3) units which encrust and penetrate the cellulose cell walls of higher plants thus contributing to their mechanical strength and rigidity—a sort of biological reinforced concrete. Lignification of plant cells is routinely detected by treatment with phloroglucinol and concentrated hydrochloric acid which yields a characteristic cherry-red colour (Weisner reaction); this reaction is due to the presence of coniferyl aldehyde linked to the remainder of the lignin polymer through a phenolic hydroxyl group. The solubilization, and removal, of lignins is a major operation in the manufacture of wood pulp for papermaking and is achieved by boiling wood chips in sodium bisulphite solution (for 'sulphite' pulp) or in an alkaline mixture of sodium sulphate and sodium sulphide (for the stronger 'kraft' pulp). These treatments break down the lignin polymer but as yet no economic use has been found for the millions of tons of lignaceous waste that are produced annually.

Lignin comprises some 22–34% of the total solid material of wood but because of its resistant nature it is difficult to be sure that lignin fractions extracted by various techniques represent anything like native lignin. The usual process is to extract finely ground wood for several days with an organic solvent such as cold ethanol or dioxane to eventually obtain a soluble, carbohydrate-free, material commonly called 'protolignin' or 'Braun's lignin'. Another technique is to utilize a 'brown-rot' fungus, such as *Lentinus lepidus*, which digests the polysaccharide material of the cell wall but which, hopefully, leaves the lignin unattacked. By contrast the less common 'white-rot' fungi attack the lignin leaving the cellulosic material.

For some time it has been known that the lignins from mono-cotyledons, dicotyledons and gymnosperms showed slight structural difference from each other and this has been exemplified by the fact that the different lignins yield different aromatic aldehydes when subjected to oxidation by alkaline nitrobenzene. These differences are summarized below.

	p-Hydroxy-benzaldehyde	Vanillin	Syringaldehyde
Gymnosperms	−	+	−
Monocotyledons	+	+	−
Dicotyledons	−	+	+

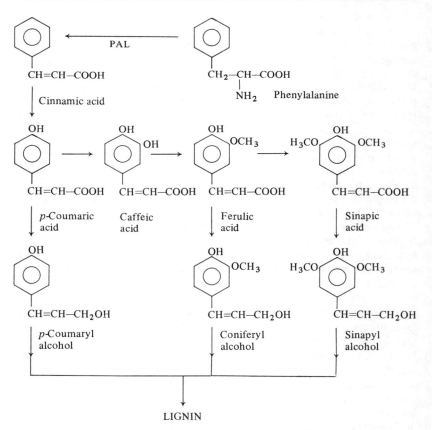

Fig. 4-1   Formation of lignin precursors. N.B. These compounds do not usually occur free but in bound or esterified forms

The above results suggest that lignin is made of phenylpropane (C_6–C_3) units and tell also what substitution patterns occur on the benzene rings—for example gymnosperm lignin is made up of C_6–C_3 units whose aromatic ring carries a *para*-hydroxyl and *meta*-methoxy group.

The immediate precursors of lignin appear to be the various substituted cinnamyl alcohols formed from the appropriate phenolic acids as shown in Fig. 4-1. However at some early stage in this pathway the plant conjugates the cinnamic acids with sugars or other compounds and this has the valuable property of rendering them less susceptible to oxidation by diphenol oxidases. However β-glycosidases have been detected in the cells adjacent to the cambial zone where lignification first takes place and these enzymes have been shown to hydrolyse coniferin (coniferyl alcohol-4-β-glucoside) to yield coniferyl alcohol which was subsequently incorporated into lignin.

Pioneer work by Freudenberg in 1953 showed that the laccase or peroxidase enzymes, which are common in plants, could oxidize a mixture of *p*-coumaryl alcohol, coniferyl alcohol and sinapyl alcohol to yield a

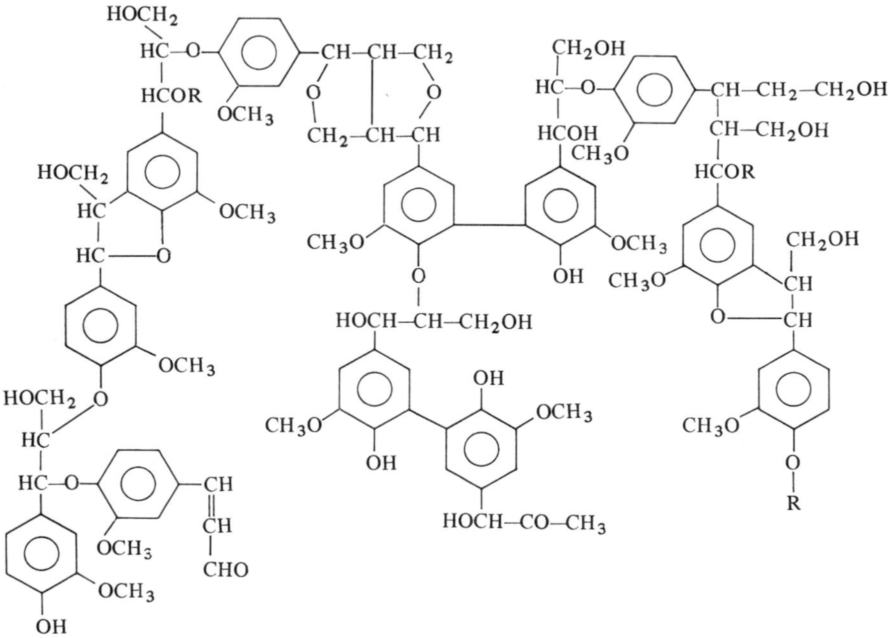

Fig. 4-2 Structural elements of part of a molecule of guaiacyl lignin. R represents the rest of lignin molecule. This is not a structural formula in the true sense, but rather a representation of linkages believed to exist in spruce lignin, and should not be interpreted quantitatively. (From Neish, 1965.)

lignin-like polymer. Freudenberg later isolated a mixture of dimers such as dehydrodiconiferyl alcohol, pinoresorcinol and guaiacyl-glycerol-β-coniferyl alcohol from this reaction system. These dimers were present as racemic mixtures which suggested that the oxidative coupling reaction involved a free radical mechanism.

On the basis of the structure and reactions of these biosynthetic intermediates Freudenberg has suggested the complex arrangement shown in Fig. 4-2 as a possible structure for lignin. The apparent lack of optical activity of lignin suggests that its optically active precursors are polymerized in a random manner allowing complex interchain and sidechain condensations to take place. (For a review see Freudenberg and Neish, 1968.)

At this point is is interesting to consider briefly the metabolism of phenyl-propanoid compounds and the evolution of plants. As plants evolved they were faced with problems of waste disposal and it is thought that many secondary plant metabolites may have been formed as part of a sort of chemical waste disposal system. Neish has suggested that lignin formation may be the result of mutations which brought about the polymerization reactions of phenolic waste materials and the deposition of this material in the cell wall matrix. This in turn conferred greater structural rigidity which was advantageous for the development of large terrestrial plants. Thus a secondary metabolite became a vital part of the plant and no longer just an inconvenient waste product.

4.2 Tannins

Tannins comprise a heterogenous group of plant polyphenols which share the common property of being able to combine with skin proteins in such a way as to render them resistant to putrefaction or, in other words, to 'tan' them into leather. More specifically tannins are high molecular weight compounds (M.W. 500–5000) containing sufficient phenolic hydroxyl groups to permit the formation of stable cross-links with proteins and as a result of this cross-linking enzymes may be inhibited. Tannins usually give rise to a dry, puckery, astringent sensation in the mouth. Because of these protein-binding properties tannins are of considerable importance in food processing, fruit-ripening and the manufacture of tea, cocoa and wine. Virtually nothing is known of the role of tannins in the life of the plant.

Tannins are classified into two broad groups: the hydrolysable and the condensed or non-hydrolysable tannins. The hydrolysable tannins are usually compounds containing a central core of glucose or other poly-hydric alcohol esterified with gallic acid (gallotannins) or hexahydroxy-diphenic acid (ellagitannins).

Condensed tannins are mostly flavolans or polymers of flavan-3-ols (catechins) and/or flavan 3:4-diols (leucoanthocyanidins). They are more resistant to breakdown and typically show a tendency to polymerize in

Gallic acid   Hexahydroxydiphenic acid

hot acid solution to yield reddish-brown coloured products; this is the well-known 'phlobaphene' reaction and is used as a test for leucoanthocyanidins.

Catechin   Leuco-anthocyanidin

The occurrence of tannins, in plants, particularly the leucoanthocyanidins, is usually associated with the woody habit and they are rarely found in herbaceous angiosperms. Their presence in trees and woody shrubs serves to render the material bitter or astringent and therefore less palatable to grazing animals. In fact tannins have been shown to depress digestion in the rumen of cattle and the digestibility of a feed appears to be correlated with its tannin content.

5 Degradation of Phenolic Compounds by Micro-organisms

We have already seen that plants possess the ability to synthesize a vast array of phenolic compounds most of which appear to be relatively inert as far as degradative processes in the plant are concerned. In fact if phenols such as quinol or *o*-chlorophenol are administered to higher plants, the plant does not metabolize them but detoxifies them by converting them to their glycosides. Metabolically speaking phenolics would seem to be at the end of the road as far as the plant is concerned and apart from the work of Zaprometov (1959), who showed that tea shoots fed ^{14}C-labelled catechin slowly released labelled CO_2, there is little evidence that higher plants can cleave aromatic rings or use phenolics as food reserves.

By contrast, and fortunately for mankind, many species of micro-organisms are able to degrade aromatic compounds thus releasing vast amounts of carbon which otherwise would be locked away in plant secondary metabolites such as lignin. In addition benificent micro-organisms are able to elaborate enzymes which break down many man-made compounds such as pesticides and detergents which contain aromatic systems. The microbes which perform these degradations are usually soil organisms and include members of such species as *Pseudomonas, Moraxella, Achromobacter* and fungi such as *Aspergillus* and *Penicillium*.

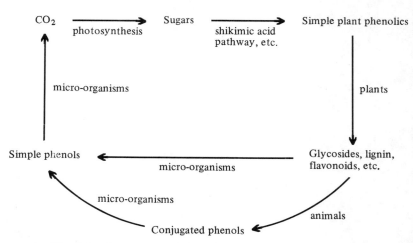

Fig. 5-1　The carbon cycle with respect to plant phenolics

Thus we may construct a special carbon cycle (Fig. 5-1) involving plant phenolics.

5.1   The microbial degradation of phenols: simultaneous adaptation

From the point of view of the organic chemist the benzene nucleus is a relatively stable structure and requires quite drastic conditions before it is attacked, yet micro-organisms operating under physiological conditions of pH (4-8) and temperature ($10\text{-}30°$) are able to open the aromatic ring by the introduction of oxygen. However, before the benzene nucleus can be split it must bear two adjacent or *ortho*-hydroxy groups so that catechol or its relatives are key intermediates in the microbial decomposition of phenolics. Frequently it is necessary for any side-chain substituents to be broken down before ring fission may take place and it was the pioneer studies of this problem by Stanier which led to the development of the technique of 'simultaneous adaptation' or 'sequential induction' which has proved to be of great value in studies of microbial metabolism.

The general idea behind this technique is best illustrated by an example. Suppose bacteria growing in a simple nutrient broth are harvested and placed in a mineral salts medium containing compound A as sole source of carbon, then a time-lag is observed before they can metabolize compound A. This delay is caused because the cells must first synthesize the battery of enzymes necessary to break down compound A and its metabolic intermediates via the sequence:

$$A \rightarrow B \rightarrow C \rightarrow D \rightarrow \text{Krebs TCA cycle}$$

Thus bacteria adapted to grow on A are now adapted to grow on compounds B, C and D and furthermore we would expect these compounds to be oxidized at rates similar to that observed for A. In this way growth on compound A appears to have 'sequentially induced' the formation of the sequence of enzymes necessary for the metabolic pathway $A \rightarrow \rightarrow D$ *et seq.* However, two important limitations must be borne in mind when interpreting this type of experiment. Firstly, a substance may be an intermediate in the pathway yet, because of permeability barriers, be unable to penetrate whole cells. Secondly, a substance may be oxidized by cells yet it is not an intermediate in the proposed pathway; however, in this case it will also be oxidized by unadapted cells.

In recent years extensions of the above approach has yielded valuable information about the mechanisms by which the biosynthesis of enzymes in a particular metabolic pathway may be controlled and it now appears that whole sub-sequences of a pathway may be co-ordinately repressed by an inducer further down the metabolic pathway.

Using the above approach together with radio-tracer studies, various workers have shown that a great many substituted phenolic compounds

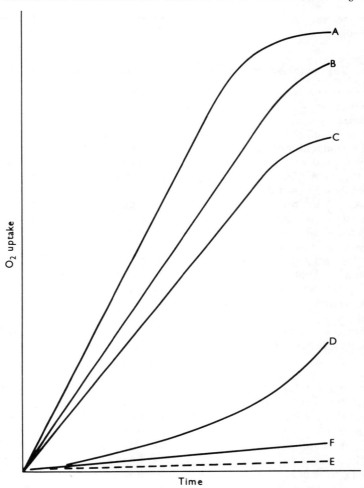

Fig. 5-2 Simultaneous adaptation (or 'sequential induction'). Micro-organisms were grown on compound A as carbon source and then tested for their ability to oxidize B, C, D and F; only B and C are likely to be intermediates in the breakdown of A. Line E is a measure of endogenous respiration in the absence of substrate

have their side-chains degraded to yield catechol-like compounds and these data are summarized in Fig. 5-3. Note that catechol and protocatechuic acid are key intermediates because these compounds contain the *ortho*-dihydroxy groups which are an essential prerequisite for ring fission. It may be mentioned in passing that the introduction of a second phenolic hydroxy group is mediated by a 'mixed function oxidase' similar to the 'cresolase' activity of higher plant diphenol oxidases.

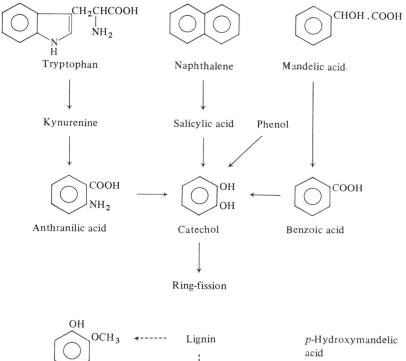

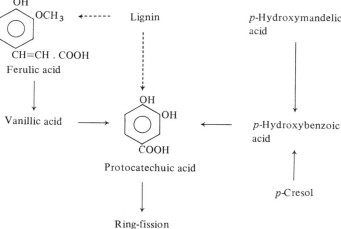

Fig. 5-3   The breakdown of phenols by micro-organisms: the central role of catechol and protocatechuic acid

5.2   The *ortho*-fission pathway for dihydroxyphenols

The pathways by which catechol and protocatechuic acid are oxidized by many species of *Pseudomonas* and other soil bacteria are summarized in

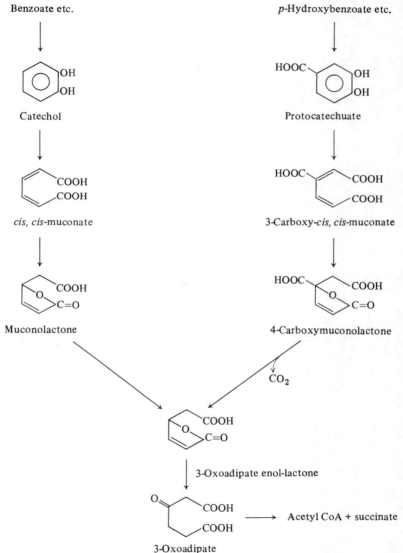

Fig. 5-4   The *ortho*-fission pathway for the breakdown of catechol and protocatechuic acid by bacteria

Fig. 5-4. It will be seen that both substances are oxidized by a parallel series of reactions catalysed by distinct sequences of enzymes until a common intermediate is reached at 3-oxoadipic-enol-lactone from whence further reactions yield succinate and acetyl coenzyme A which may then enter the Krebs tricarboxylic acid cycle.

Detailed studies by Dagley (1971) and other workers of these reactions have provided valuable information on the mechanisms by which micro-organisms regulate their metabolic processes.

5.3   The *meta*-fission pathway for catechol

The *ortho*-fission pathway involved cleavage of the benzene ring between adjacent hydroxy groups but in 1959 Dagley and Stopher found that certain micro-organisms such as *Pseudomonas testosteroni* or *Nocardia restrictus* opened the ring *meta*- to the two *ortho*-hydroxy

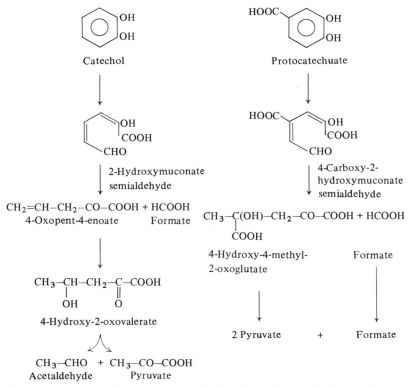

Fig. 5-5   The breakdown of catechol and protocatechuic acid by the *meta*-fission pathway

groups. By contrast with the *ortho*-fission route the first product of the *meta*-pathway is a hydroxy-muconic semialdehyde and this ultimately yields pyruvate together with an aldehyde whose structure depends upon the phenol being oxidized. Unlike the *ortho*-fission route 3-oxo-adipic acid is not an intermediate in the *meta*-fission pathway.

5.4   The breakdown of lignin

The previous discussion will have served to show that micro-organisms possess remarkable facilities to break down complex organic compounds and it is pertinent to consider the breakdown of lignin since this material represents a vast storehouse of polymerized phenolic material in plants. Moreover it plays a part in the formation of humus in the soil.

As discussed in Section 4-1 lignin is considered to be a complex three-dimensional polymer of phenylpropane units and when acted upon by various soil micro-organisms these are released as coniferyl alcohol, ferulic acid and other similar products which are subsequently oxidized via one or other of the catechol pathways as outlined above.

The initial microbial attack on wood and lignin is throught to be due to various fungi which are usually divided into two groups: the white-rot fungi and the brown-rot fungi. The former include many common 'bracket' fungi such as *Polyporous* sp and this group is capable of utilizing both lignin and cellulose as a source of carbon for growth. By contrast the brown-rot fungi preferentially attack the cellulosic components of wood and the remaining brown material is sometimes used as a source of lignin for chemical studies. A number of workers (ref. Kirk 1971 for a review) have shown that when white-rot fungi were grown on sawdust they yielded a variety of phenolic lignin constituents including such compounds as coniferaldehyde, vanillin, syringaldehyde together with the related vanillic and syringic acids and these compounds were subsequently oxidized by soil bacteria. For example a soil *Pseudomonad* was found to oxidize ferulic acid to yield vanillic acid which in turn was oxidized to protocatechuic acid and thence to CO_2 (See Fig 5-3 *et seq.*). However, an *Achromobacter* sp cleaved the aromatic nucleus of hydroxycinnamic and caffeic acids before degradation by the *meta*-fission pathway.

5.5   Degradation of chlorinated phenols

In view of current concern with the environment it is of interest to mention briefly how soil micro-organisms cope with some of the modern chlorine-substituted phenolics that ultimately find their way into the soil.

For example, the herbicide 2:4-dichlorophenoxy acetic acid (2:4-D) is converted to 2:4-dichlorophenol which is subsequently cleaved to 2-chloromuconic acid. More recently Horvath (1972), introduced the concept of co-metabolism having shown that although chloro-substituted phenols cannot support microbial growth as sole source of C they may be degraded by enzymes induced when the organisms grow on the related unsubstituted compounds. Thus a complex organic compound may be oxidized without actually supporting microbial growth.

The significance of this phenomenon in the life of the soil remains to be evaluated.

6 Experimental Work with Plant Phenolics

There are a number of investigations in this field that should make interesting projects for sixth and seventh form students yet are not too demanding of specialized equipment or reagents. It is hoped that the following experiments may stimulate students' interest in plant phenolics.

6.1 Enzymic browning: a study of diphenol oxidase

The significance of phenolic compounds in fruits was discussed in Section 3.2 and this topic can form the basis of a study of the properties of diphenol oxidases or a consideration of the problem of enzymic browning in food processing.

Diphenol oxidase activity in fruits and vegetables may be demonstrated by applying a few drops of 0.1% catechol solution to a freshly cut surface and observing the formation of a brown-coloured zone. This simple technique can be used for many of the subsequent experiments. However, the experiments are better carried out using a soluble enzyme preparation prepared by grinding about 20 g of potato tissue with sand in a pestle and mortar in the presence of 100 ml 0.05 M-sodium fluoride solution (caution POISON). The resultant slurry is quickly filtered through fine muslin and the liquid used as a source of crude potato diphenol oxidase (DPO) in the following experiments.

(a) Enzyme action

Set up test tubes containing the following reactants and shake the tubes at 5 minute intervals: (why?).

Tube	Enzyme	Water	0.01 M-Catechol
1	2.0 ml	1.0	− (control)
2	2.0 ml	−	1.0
3	1.0 ml	1.0	2.0
4	1.0 ml	−	2.0
5	0.5 ml	1.5	1.0

Follow the time course of the reaction at 5 minute intervals using an arbitary 0–5 scale to record colour formation. Plot your results on a graph and comment on the effect of altering the amounts of enzyme and substrate. Does product inhibition occur in this experiment?

Devise simple modifications to the above procedure to demonstrate (i) that an enzyme is involved and (ii) that O_2 is required for the reaction.

(b) Effect of temperature

Set up a series of control and experimental tubes (1 and 2) as for experiment (a) above but incubate each pair at different temperatures ranging from ice-water to a boiling water bath. Record colour changes and graph your results. At what temperature does DPO activity fall off? Why might this sort of experiment be of interest to food technologists?

(c) Effect of pH

Prepare a range of phosphate-citrate buffers covering pH 4.0–8.0 and add 1 ml of each buffer to 1 ml of DPO followed by 1 ml of catechol solution. Follow the reaction as before and determine the optimum pH of potato DPO.

(d) Substrate specificity

Investigate the rate of reaction and colour produced when different phenolic substrates (all 0.01 M) are used in place of catechol (Tube 2). The following phenols might be tested; phenol, resorcinol, quinol (hydro-quinone), pyrogallol and p-phenylene diamine. Try to correlate structure and DPO activity.

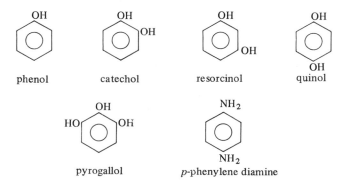

phenol   catechol   resorcinol   quinol

pyrogallol   p-phenylene diamine

(e) Action of enzyme inhibitors

Using the basic catechol/DPO assay system investigate the effect of the addition of various concentrations (try 10^{-4}–10^{-2} M) of the following inhibitor substances:

(i) Ascorbic acid, $NaHSO_3$, cysteine: these act as anti-oxidants.
(ii) Thiourea, phenyl-thiourea, sodium diethyldithio-carbamate (DIECA): these are all Cu-reagents.
(iii) Chloride ions (as NaCl).

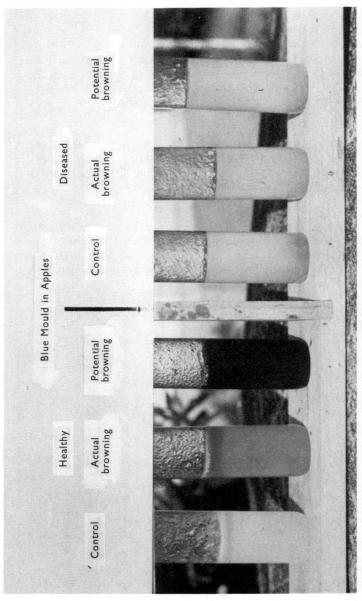

Fig. 6-1 Actual and potential browning of apples. The tubes contained homogenized apple pulp plus the following additions; (a) Control; thiourea added to inhibit diphenol oxidase; (b) Actual browning; no additions, became light brown; (c) Potential browning; catechol added, became dark brown–black. This picture is taken from the author's research into the biochemistry of plant disease (Walker, 1969). The tubes on the right hand side of the picture contained pulp from apple tissue infected with *Penicillium expansum* and failed to brown because of the presence of enzyme inhibitors secreted by the invading fungus.

Which of these methods might be used to control enzymic browning in fruit-processing operations?

If a Warburg apparatus or an O_2-electrode is available the above experiments may be readily adapted to yield quantitative results and the Michaelis constant (K_m) of the enzyme might be determined.

An alternative approach more related to problems encountered in food processing would be to compare the levels of actual and potential browning of different varieties of apple, pear or potato. Actual browning is a measure of the amount of browning given by the pulped tissue without added substrate and it is a measure of the level of natural substrates present. Potential browning measures the total enzymic browning when an excess of substrate (catechol) is added.

6.2   A study of laccases in wood-rotting fungi

As long ago as 1928 Bavendamm showed that many wood-rotting fungi secreted extracellular p-diphenol oxidases or laccases which could be detected by growing the fungus on a medium containing gallic or tannic

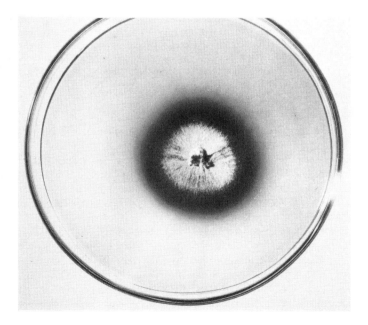

Fig. 6-2   Bavendamm reaction. A culture of *Coriolus (Polystictus) versi-color* grown on potato glucose agar containing 10^{-3} M tannic acid. The dark brown zone around the fungal colony indicates the presence of an extracellular laccase

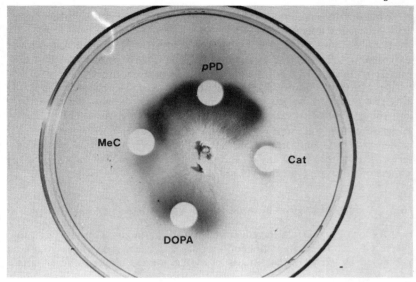

Fig. 6-3   Detection of extracellular diphenol oxidases. In this experiment the fungus was grown on potato glucose agar and various phenolic substrates placed in the small wells cut into the agar. Development of colour indicates a positive test. Key: MeC = 4-Methylcatechol: brown. DOPA = 3:4-Dihydroxyphenylalanine: dark brown. pPD = p-Phenylene diamine: purple. Cat = Catechin: orange

acid. Laccase formation is revealed by the formation of a brown zone of oxidized substrate around the mycelium.

To carry out this test prepare potato glucose agar or malt extract agar plates containing 10^{-3} M-gallic acid (0.019%) or tannic acid (0.17%) and inoculate them with a selection of fungi from decaying timber, leaf mould, diseased fruit and any other likely source. Try to correlate laccase activity with habitat and/or pathogenecity.

Fungi which yield positive Bavendamm tests are good potential sources of diphenol oxidase for advanced studies. They should be grown on a simple liquid medium containing 3% malt extract and the culture filtrate tested for DPO activity at 2- or 3-day intervals.

It may also be of interest to test for extracellular laccases by growing the fungi on normal malt extract agar and when part grown placing drops of 0.1% solutions of various phenolic substrates around the edge of the growing mycelium. This study may yield interesting information on substrate specificity differences between different fungi and this approach has been investigated as an aid to the identification of wood-decaying fungi (Käärik 1965).

6.3   The anthocyanin pigments of plants

The cell sap of higher plants frequently contains water soluble pigments in the form of red or blue anthocyanins together with pale yellow or ivory coloured flavonoids and the structure of these compounds has been reviewed in Chapter 1. However these pigments should be distinguished from the lipid-soluble orange or red carotenoid pigments which are located in the plastids, although both groups of pigments contribute to the final colour of a leaf or flower. Moreover the final colour of the anthocyanins in the cell sap may be influenced by pH, metal chelation and other factors and some of these effects may be readily demonstrated.

Anthocyanin and flavonoid pigments are also of great interest to plant geneticists and systematists because it is often possible to correlate phenotypic differences among closely related species or to follow lines of evolution of plants within a genus and it is hoped that the following experiments will serve to guide students to design their own investigations into this fascinating multi-discipline area of research.

The main stages in a study of anthocyanin pigments usually involve extraction into methanol, hydrolysis to determine the aglycone and sugar moieties and identification of the anthocyanidin by spectroscopy and other methods. Using this approach students and teachers may find the following lines of investigation profitable:

(a) A comparison of anthocyanin pigments in flowers, fruits and leaves from the same or different plants.
(b) The distribution of pigments for different varieties within a species or species within a genus. Suitable material might be varieties of *Streptocarpus, Primula, Lathyrus, Antirrhinum, Viola,* etc.
(c) The effect of physiological factors such as light, pH and mineral nutrition upon pigmentation, for example an investigation of pigments in blue and red *Hydrangea* flowers. Alternatively, the effect of metal chelation upon anthocyanins might be studied since this problem affects food processing, wine-making and similar processes.

6.4   Analytical techniques for anthocyanins and flavonoids

6.4.1   Extraction procedures

Anthocyanins are normally extracted from fresh or dried plant material by grinding about 1 g of fresh material or 0.1 g dried material in 0.5–2.0 ml of methanol (MeOH) containing 1% concentrated HCl; a little fine sand may assist the extraction. The supernatant solution is then filtered through Whatman No. 4 paper and extracted with petroleum ether to remove chlorophyll and carotenoids. The remaining methanolic extract will contain the anthocyanins and flavonoids and will be used for all subsequent experiments.

N.B. Solutions of anthocyanins are unstable; they should be kept acid and stored away from light in a cool place.

6.4.2   Hydrolysis of anthocyanins

Examination of the crude extract prepared above may reveal the presence of many different glycosides of the same parent anthocyanidin aglycone so in order to simplify the problem, it is advisable to hydrolyze the anthocyanins by treating an aliquot of the crude extract with an equal volume of 2 N-HCl. Heat the mixture on a boiling water bath for 30–60 minutes to ensure complete hydrolysis.

It may be of interest to follow the time course of hydrolysis by taking samples at 5–10 minutes intervals and applying them to a chromatogram. Thus hydrolysis of an anthocyanin with a tri-glycoside side-chain will yield successively the di- and mono-glycosides and finally the aglycone; in this way the substitution pattern may be investigated.

When hydrolysis is complete the anthocyanidins may be extracted into a small quantity of amyl alcohol (AmOH) from which they may then be precipitated by the careful addition of an excess of benzene. However, the AmOH extract will serve for most subsequent operations.

In more sophisticated studies each anthocyanin would be separated and then individually hydrolyzed to establish the identity of its aglycone and number of sugar residues. Likewise the sugars would be identified by conventional chromatographic procedures.

6.4.3   Chromatographic procedures

Anthocyanins and anthocyanidins are best separated by chromatography on paper or cellulose thin-layers and it is assumed that readers are familiar with the basic techniques involved. The author has found the solvent systems listed in Table 6-1 to be adequate for most purposes.

Table 6-1   Solvents for chromatography of anthocyanins

Name	Composition	Approximate time (h) for solvent front to run 15 cm.
BAW	n-Butanol, acetic acid, water (40:10:22)	5
HAc	2% (v/v) Acetic acid	$1-1\frac{1}{2}$
HCl	1% HCl (3 ml conc HCl + 97 ml water)	$1-1\frac{1}{2}$
Forestal*	Acetic acid, conc HCl, water (30:3:10)	3–6

* For Anthocyanidins

When the chromatogram is developed it should be dried and the location of visible spots outlined in pencil. It should then be examined under long wavelength UV light to locate any fluorescent flavonoid compounds.

If larger quantities of pigments are required for standards or spectroscopy then the crude material from the AmOH or MeOH extracts should be run on Whatman No. 3MM paper.

Two dimensional chromatograms are often useful for resolving complex mixtures of pigments. In this case a concentrated sample of the mixture is placed in the corner of the chromatogram and the paper or plate developed in BAW. It is then allowed to dry, rotated through 90° developed in a suitable second solvent such as HAc or Forestal (for anthocyanidins).

6.4.4   Identification of anthocyanidins

The separated anthocyanidins may be tentatively identified from their colours and the R_F values given in Table 6-2, but these can serve only as a rough guide. More certain identification can only be gained by comparing R_F values of the samples with those of authentic specimens chromatographed under the same conditions together with comparison of absorption spectra. In view of this it is useful to prepare reference samples of various anthocyanins and anthocyanidins from suitable specimens known to possess only one major pigment and comprehensive compilations of suitable sources of anthocyanins are given in the reference texts. However, a brief list of sources which yield anthocyanidins after hydrolysis is included here.

Pelargonidin: strawberry fruit, pomegranate flowers, pelargonium flowers
Cyanidin: rose petals, copper beech leaves, apple fruit skins, blackberry or elderberry fruits, red cabbage
Delphinidin: delphinium flowers, hydrangea flowers, bluebell flowers
Malvidin: grape (skins), cyclamen flowers
Peonidin: peony flowers
Petunidin: petunia flowers, flowers and berries of deadly nightshade

For those people lucky enough to have access to spectrophotometers, measurements of the absorption spectra of anthocyanidins in 1% HCl in MeOH is a valuable means of confirming their identity since the λ_{max} is characteristic of each pigment. Furthermore, those anthocyanidins which possess free *ortho*-hydroxy group on their B-rings (cyanidin, delphinidin, and petunidin) show bathochromic shifts of their λ_{max} when a drop of $AlCl_3$ in MeOH is added to the solution.

The following books, additional to those cited in the reference section, provide useful information on practical aspects of work with anthocyanins.

Dunn, A. and Arditti, J. (1968). *Experimental Physiology*. Holt, Rinehart and Winstone Inc. U.S.A., p. 138-154.
Clevenger, S. Flower Pigments. *Scientific American*. June, 1964 (reprint no. 186). W. H. Freeman and Co., U.S.A.

Table 6-2 Identification of common anthocyanidin pigments

| Pigment | Colour | R_F value on Whatman No. 1 paper Solvent* | | | λ_{max} in MeOH–HCl (nm) | $\Delta\lambda_{max}$ with $AlCl_3$ (nm) |
		Forestal	BAW	Formic		
Pelargonidin	Scarlet-red	68	80	33	520	0
Cyanidin	Magenta	49	68	22	535	+18
Peonidin	Magenta	63	71	30	532	0
Delphinidin	Purple-blue	32	42	13	546	+23
Petunidin	Purple-blue	46	52	20	543	+14
Malvidin	Purple	60	58	27	542	0

* Forestal: acetic acid, HCl, water (30:3:10); BAW: n-Butanol; acetic acid, water (40, 10, 22); Formic: Formic acid, HCl, water (5:2:3)

Further Reading and References

General texts

Geismann, T. A. (1962). *The Chemistry of Flavonoid Compounds.* Pergamon Press, Oxford.

Harborne, J. B. (1964). *Biochemistry of Phenolic Compounds.* Academic Press, London and New York.

Harborne, J. B. (1967). *Comparative Biochemistry of the Flavonoids.* Academic Press, London and New York.

Ribereau-Gayon, P. (1972). *Plant Phenolics.* Oliver & Boyd, Edinburgh.

Swain, T. (1963). *Chemical Plant Taxonomy.* Academic Press, London and New York.

Specialist articles

Brown, S. A. (1966). Lignins. *A. Rev. Pl. Physiol.,* **17**, 223–244.

Dagley, S. (1965). Degradation of the Benzene Nucleus by Bacteria. *Science Progress,* **53**, 381–392.

Dagley, S. (1971). Catabolism of Aromatic Compounds by Micro-Organisms. *Adv. in Microbial Physiol.,* **6**, 1–46.

Ingham, J. L. (1972). Phytoalexins and Other Natural Products as Factors in Plant Disease Resistance. *Bot. Rev.,* **38**, 343–424.

Kirk, T. K. (1971). Effects of Micro-organisms on Lignin. *A. Rev. Phytopathol.,* **9**, 185–210.

Levin, D. A. (1971). Plant Phenolics: An Ecological Approach. *American Naturalist,* **105**, 157–181.

Muller, C. H. and Chang-Hung Chou (1972). Phytotoxins: An Ecological Phase of Phytochemistry. In *Phytochemical Ecology.* Edited by J. B. Harborne. Academic Press, London and New York.

Pridham, J. B. (1965). Low Molecular Weight Phenols in Higher Plants. *A. Rev. Pl. Physiol.,* **16**, 13–36.

Singleton, V. L. and Esau, P. (1969). *Phenolic Substances in Grapes and Wine and their Significance.* Academic Press, London and New York.

Stafford, H. A. (1974). The Metabolism of Aromatic Compounds. *A. Rev. Pl. Physiol.,* **25**, 459–486.

Woodcock, D. (1964). Microbial Degradation of Synthetic Compounds. *A. Rev. Phytopathol.,* **2**, 321–340.

Books on methodology

Harborne, J. B. (1973). *Phytochemical Methods.* Chapman and Hall, London.

Mabry, T. J., Markham, K. R. and Thomas, M. B. (1970). *The Systematic Identification of Flavonoids*. Springer-Verlag, New York.
Pridham, J. B. (1964). *Methods in Polyphenol Chemistry*. Pergamon Press, Oxford.

References

Bate-Smith, E. C. (1956). *Sci. Proc. R. Dublin Soc.*, **27**, 165.
Bavendamm, W. (1928). *Z. Pflzkrankh. Pflzschutz.*, **28**, 257.
Camm, E. L. and Towers, G. H. N. (1973). *Phytochem.*, **12**, 1-13.
Challice, J. S. and Williams, A. H. (1966). *J. Chromatog.*, **21**, 357-362.
Conn, E. E. (1964). Enzymology of Phenolic Biosynthesis. In *Biochemistry of Phenolic Compounds*. Edited by J. B. Harborne. Academic Press, London and New York.
Dagley, S. and Stopher, D. A. (1959). *Biochem. J.*, **73**, 16P.
Davis, B. D. (1955). *Adv. Enzymol.*, **16**, 247.
Freudenberg, K. and Neish, A. C. (1968). *Constitution and Biosynthesis of Lignin*. Springer-Verlag, Berlin.
Finkle, P. J. and Nelson, R. F. (1963). *Biochem. Biophys. Acta*, **78**, 747.
Grisebach, H. (1965). Biosynthesis of Flavonoids. In *Chemistry and Biochemistry of Plant Pigments*. Edited by T. W. Goodwin. Academic Press, London and New York.
Grisebach, H. (1967). *Biosynthetic Patterns in Micro-organisms and Higher Plants*. John Wiley and Sons, New York.
Harborne, J. B. and Simmonds, N. W. (1964). The Natural Distribution of the phenolic aglycones. In *Biochemistry of Phenolic Compounds*. Edited by J. B. Harborne. Academic Press, London and New York.
Horvath, R. S. (1972). *Bact. Rev.*, **36**, 146-155.
Käärik, A. (1965). The identification of the mycelia of wood-decay fungi by their oxidation reactions with phenolic compounds. Studia Forestalia Seucica. Nr 31.
Kojima, M. and Uritani, I. (1972). *Plant and Cell Physiol.*, **13**, 311-319.
Lee, S. and Aronoff, S. (1967). *Science*, **158**, 798-799.
Levy, C. C. and Zucker, M. (1960). *J. Biol. Chem.*, **235**, 2418-2425.
Muller, K. O. and Borger, H. (1940). *Arb. biol. Abt. (Ansl.-Reichsanst), Berl.*, **23**, 189.
Neish, A. C. (1965). Coumarins, phenylpropanes and lignin. In *Plant Biochemistry*. Edited by J. Bonner and J. E. Varrer. Academic Press, London and New York.
Rajarnatam, J. A., Lowry, J. B., Avadhani, P. N. and Corley, R. H. V. (1971). *Science*, **172**, 1142-1143.
Reznik, H. (1956). *S. B. Heidelberg. Akad. Wiss.* 125.
Satô, M. (1966). *Phytochem.*, **5**, 385-389.
Swain, T. and Bate-Smith, E. C. (1962). Flavonoid Compounds. In *Comparative Biochemistry* Vol. III. Edited by M. Florkin and H. S. Mason. Academic Press, London and New York.
Tomaszewski, M. (1960). *Bull. Acad. polon. Sci.*, **8**, 61.

Underhill, E. W., Watkin, J. E. and Neish, A. C. (1957). *Canad. J. Biochem. Physiol.*, **35**, 219.
Walker, J. R. L. (1964). *Aust. J. Biol. Sci.*, **17**, 360–371.
Walker, J. R. L. (1969). *Phytochem.*, **8**, 561–566.
Wong, E. (1968). *Phytochem.*, **7**, 1751–1758.
Zaprometov, M. N. (1959). *Dokl. Akad. Nauk. S.S.S.R.*, **125**, 1359.